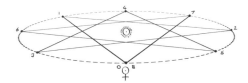

Climb up a hill at midday on your birthday every year and look at the Sun. Each year Venus will be three-eighths of the way further round the Sun, drawing a perfect octagram over eight years.

First published 2001AD
This revised edition
© Wooden Books Ltd 2012

Published by Wooden Books Ltd.
8A Market Place, Glastonbury, Somerset

British Library Cataloguing in Publication Data
Martineau, J. S.
A Little Book of Coincidence

ISBN 1 904263 05 4

Printed and bound in China
by Shanghai iPrinting Co., Ltd.
100% recycled papers.

A LITTLE BOOK OF
COINCIDENCE
IN THE SOLAR SYSTEM

written and illustrated by

John Martineau

*Thanks to the many friends, colleagues and other nutters who have
contributed to this project over the years. Please keep crunching the
numbers and sending in the coincidences. Rich patrons may send funds.*

Note: Accuracies in this book are shown as simple percentages.

*Early visions of an infinite universe of solar systems hinted at repeated structures
like galaxies and parallel universes. From Thomas Wright's "The Cosmos", 1750.*

CONTENTS

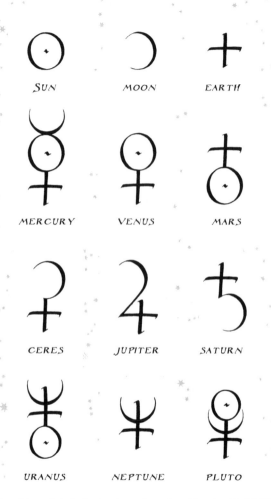

SUN

MOON

EARTH

MERCURY

VENUS

MARS

CERES

JUPITER

SATURN

URANUS

NEPTUNE

PLUTO

A useful set of glyphs for the planets drawn by calligrapher Mark Mills,
each made from Sun, Moon and Earth and used throughout this book.

INTRODUCTION

Biological life is now thought to have appeared on this planet not long after its formation. It seems that the bacterial seeds for the process may have flown in on the tail of a comet or meteor. Speculation is again rife about life under the surface of Mars, on Jupiter's icy moon Europa and indeed anywhere the sacred substance of liquid water is known to exist.

The science of the cosmos has changed immeasurably since the Greek and medieval visions of circles of planetary spheres. But with great cosmic schemes out of fashion, and with dragons and unicorns dismissed, the Earth has become a modern mystery. No modern theory exists to explain the miracle of conscious life and the numerous cosmic coincidences which surround our planet are swept under the carpet. Ancient answers to such questions invoked liberal arts like music and geometry.

This book is not just another pocket guide to our solar system, for it suggests there may be fundamental relationships between space, time and life which have not yet been understood. These days we scan the skies listening for intelligent radio signals and looking for another planet a little like our own. In recent years we have discovered that mellow solar systems like our own are very very rare indeed. Yet meanwhile, our closest planetary neighbours are making the most exquisite patterns around us, in space and in time and no scientist has yet explained why. Is it *all just a coincidence* or do the patterns perhaps explain the scientists ...

GALACTIC DUST
the well-tuned universe

There's a lot going on in the universe. We can now see as many stars within our space-time horizon-bubble as there are grains of sand on Earth. Our planet and we ourselves are made from reorganised smoky stardust, a fact long taught by ancient cultures. We now know that stardust itself is made simply from fizzballs, highly organised flickering whirlpools of light, long ago squeezed together deep inside stars. We ourselves live in between the little and the large, in a time and a place in the universe where things have condensed, crystallized, built up and settled down.

We still have no evidence as to whether conscious life is rare or common in galaxies. Just how special are we and our Earth? Funnily enough, scientists are currently puzzling over the strange fact that the whole *universe* seems special. There is *exactly* enough material in the universe to make it, and the ratios between the fundamental forces seem *specifically* tuned to produce an amazingly complex, beautiful and enduring universe. Fiddle with any bit of it, even slightly, and you get a universe of black holes, insubstantial fizzballs, or other lifeless set-ups. Is this design or coincidence?

The story of the search for order, pattern and meaning in the cosmos is very old. The planets of our solar system have long been suspected of hiding secret relationships. In antiquity students of such things pondered the *Music of the Spheres*, today they experiment with the simple precision of Kepler's, Newton's and Einstein's laws. Who knows what will come next?

YOU ARE HERE

THE SOLAR SYSTEM
spirals everywhere

Our solar system seems to have condensed from the debris of an earlier version some five billion years ago. A Sun formed in the centre and remaining materials were similarly attracted to each other to form small rocky asteroids. Lighter gases were blown out by the solar wind to condense as the four gas giants, Jupiter, Saturn, Neptune and Uranus while in the inner solar system asteroids grew into planets, the final pieces flying into place with more and more energy as the sizes grew (many still have molten cores from these collisions). Our solar system takes the form of a stable disk, a design now known to be very rare.

The plane of the solar system is tilted at roughly 30° to the plane of the galaxy so our solar system actually corkscrews its way around the arm of the milky way. The picture (*opposite above, after Windelius & Tucker*) is schematic of the motions of the four inner planets.

Another way to picture the Solar System is by thinking of space-time as a rubber sheet with the Sun as a heavy ball and planetary marbles placed on it (*opposite below, after Murchie*). This is Einstein's model of the way matter curves space-time and helps visualise the force of gravity between masses. If we flick a tiny frictionless pea onto our sheet, it could easily be captured by one of the marbles, or be spun around a few times and spat out, or settle into a fast spinning elliptical orbit halfway down any one of the worm-holes. Like a planet, the further the pea gets down the funnel, the faster it must circle to stop itself going down the tube. Also, the faster it spins the heavier it gets and the slightly slower its clocks run.

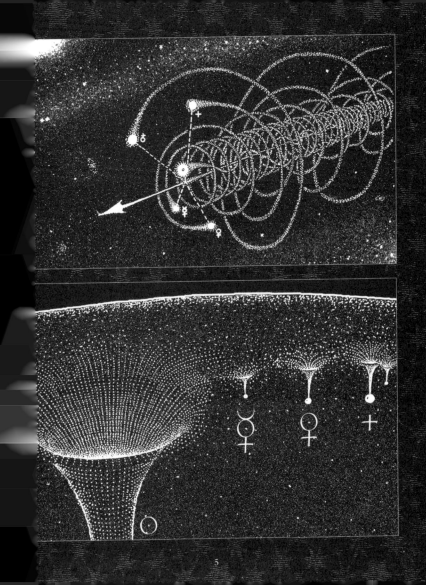

RETROGRADE MOTION
running kissing around

Ancient astronomers who watched the sky from Earth noticed that apart from the Sun and Moon there were five easily visible points of light which moved across the stars. These are the planets, which seem to move around the earth roughly following the Sun's yearly circle, the *ecliptic* or the stars of the *zodiac*. If only life was this simple! Watch planets for any length of time and, far from moving in any simple way, they lurch around like drunken bees, waltzing and whirling. As two planets pass, or kiss, each appears to the other to *retrogress* or go backwards against the stars for a certain length of time.

The diagram below shows Mercury's pattern around a tracked Sun over a year as seen from Earth (*after Schultz*), and opposite we see Cassini's 18th century sketch of the movements of Jupiter and Saturn as seen from Earth. In ancient times hugely complex systems of circles and wheels were called into play to try to mimic these planetary motions (*opposite below*), culminating in the Ptolemaic system of 39 *deferents* and *epicycles*, used to model the motions of the seven heavenly bodies over two thousand years ago.

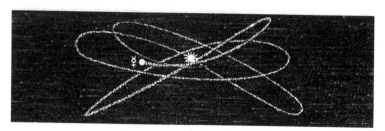

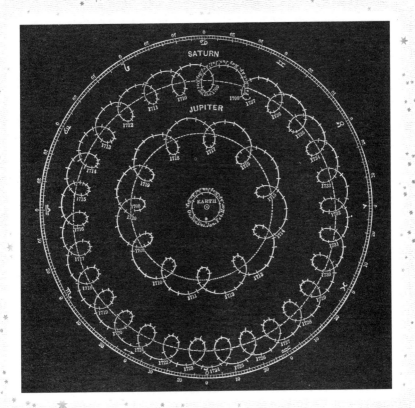

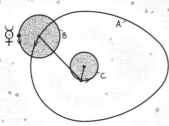

Until 400 years ago planetary motions were modelled using a 'deferent' (A) and an 'epicycle' (B). Other tricks refined the system - here a kind of crank (C) called a 'movable eccentric' produces an egg-shaped deferent for Mercury's dance.

7

THE SECRET OF SEVENS
planets, metals and days of the week

A short four hundred years ago the diagrams opposite formed the cornerstone of cosmological scientific and magical thought across the western world, as they had done for many thousands of years. Today these emblems of the seven-fold system of antiquity appear as quaint reminders of an alchemical cosmology now buried beneath newly discovered planets and physical elements.

There are seven clearly visible wandering heavenly bodies, and they may be arranged around a heptagon in order of their apparent speed against the fixed stars. The Moon appears to move fastest, followed by Mercury, Venus, the Sun, Mars, Jupiter and Saturn (*top left*). Planets were assigned to days, still clear in many languages, and the order of the days was given by the heptagram shown (*top right*). In English, older names for the planets (or gods), were used, thus *Wotan's day, Thor's day* and *Freya's day*.

In antiquity seven metals were held to correspond with the seven planets, their compounds giving rise to colour associations. Venus, for example, was associated with the greens and blues of copper carbonates. Students of alchemy would often ponder these relationships as they forged ever more subtle things. Incredibly, the ancient system also gives the *modern* order by atomic number of these metals! Follow a more open heptagram to give *Iron 26, Copper 29, Silver 49, Tin 50, Gold 79, Mercury 80 and Lead 82 (lower left after Critchlow & Hinze)*. The electrical conductivity sequence also appears round the outside starting with Lead.

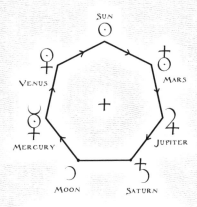

THE SEVEN HEAVENLY BODIES:
Start at the Moon and follow the arrows to give the 'Chaldean Order' of the spheres.

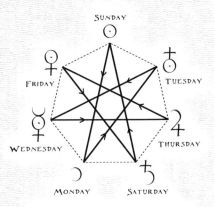

THE SEVEN DAYS OF THE WEEK:
French: Lundi, Mardi, Mercredi, Jeudi, Vendredi... follow the arrows again.

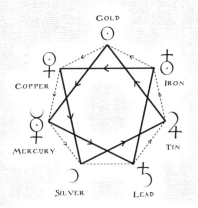

THE SEVEN METALS OF ANTIQUITY:
Start with Iron and follow the arrows to give elements of increasing atomic number.

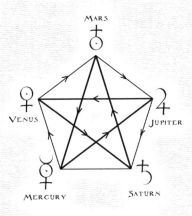

THE FIVE WANDERERS:
Start with Mercury. Moving round the pentagon increases distance from the Sun.

9

GEOCENTRIC OR HELIOCENTRIC
Earth or Sun at the centre

The extraordinary Ptolemaic world of epicycles and deferents lasted a surprisingly long time. Despite its complexity it 'saved appearances' and was also said to save souls. Ellipses were in fact studied by early Greek mathematicians such as Appollonius, and as early as 250 BC Aristarchus of Samos was proposing a system of planets orbiting the Sun. However, it was not to be, and for one and a half thousand years the Earth remained in the centre of the universe as the Ptolemaic system was handed down from the Greeks to the Arabs, and then back to the West again.

Four early systems are shown opposite (*after Koestler*), and each sphere of each diagram is to be understood as having its own attachment of epicycles and eccentrics. Copernicus, despite in 1543 placing the Sun in the centre (*top left*), remained a devout epicycle man, increasing the number of invisible wheels from the Ptolemaic 39 up to an amazing 48. In the late sixteenth century Tycho de Brahe desperately tried to keep the Earth stationary in the centre of the universe (*bottom left*), whilst an early Greek model, Herakleides', like a later version by Eriugina, attempted a compromise.

During the 1600's the tide finally turned and people began to forget that planets sometimes move backwards. A modern model of the Solar System is shown lower opposite with the planets (including an asteroid, Ceres), orbiting the Sun, each planet with an orbital 'shell'. This basic model was first conceived by Johannes Kepler in 1596 and it is to his ideas that we now turn.

Aristarchus & Copernicus

Ptolemy

Tycho de Brahe

Herakleides

Kepler

KEPLER'S VISIONS
ellipses and nested solids

Kepler noticed three things about planetary orbits. Firstly that they are ellipses (*so that a + b = constant opposite below*), with the Sun at one focus.   Secondly, that the *area* of space swept out by a planet in a given time is constant. Thirdly, that the period T of a planet relates to R, its semi-major axis (or 'average' orbit), so that T^2/R^3 is a constant throughout the entire solar system.

Looking for a geometric or musical solution to the orbits, Kepler observed that six heliocentric planets meant five intervals. The famous geometric solution he tried was to fit the five *Platonic Solids* between their spheres (*opposite, and detailed below*).

In recent years, far from diminishing Kepler's vision, Einstein's laws actually showed that the tiny space-time effects caused by Mercury's faster (and therefore heavier and time-slowed) motion when nearer to the Sun create a precessional rotation of the ellipses over thousands of years, thus reinforcing Kepler's shells.

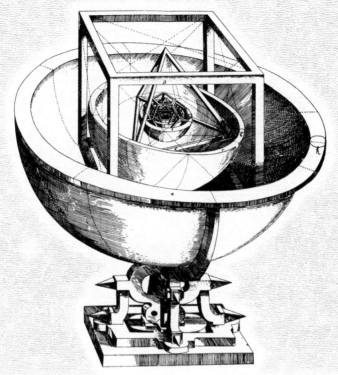

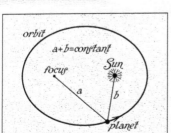

orbit

a+b=constant

focus

Sun

a

b

planet

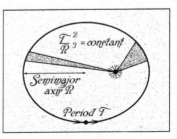

$\frac{T^2}{R^3} = constant$

Semimajor axis R

Period T

THE MUSIC OF THE SPHERES
planets playing in tune

In ancient times the seven musical notes were assigned to the seven heavenly bodies in various symbolic arrangements (*opposite top*). With his accurate data, Kepler now set about precisely calculating these long imagined *Harmoniae Mundi*. He particularly noticed that the ratios between planets' extreme angular velocities were all harmonic intervals (*opposite centre, after Godwin*). More recently, work by Molchanov has shown that the entire solar system can be viewed as a 'tuned' quantum structure, with Jupiter as the conductor of the orchestra.

Music and Geometry are close bedfellows and Weizsacker's theory of the condensation of the planets (*opposite after Murchie & Warshall*) throws yet more dappled light on to these elusive orbits. It might appear fanciful were it not for the fact that two nested pentagons (*below left*) define Mercury's shell (99.4%), the empty space between Mercury and Venus (99.2%), Earth and Mars' relative mean orbits (99.7%), and the space between Mars and Ceres (99.8%). Three nested pentagons (*below right*) define the empty space between Venus and Mars (99.6%) and also Ceres and Jupiter's mean orbits (99.6%).

ANCIENT EGYPTIAN SYSTEM

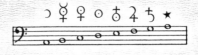

CICERO - SCIPIO'S DREAM

KEPLER'S HEAVENLY HARMONIES

Bode's Law and Synods
harmonics and rhythmic kisses

There have been numerous attempts to discover patterns in the orbits and periods of the planets. A basic logarithmic graph (*opposite top*) shows clear underlying order (*after Ovendon & Roy*).

A famous system is the 1750 *Titius & Bode Rule*: To the series 0, 3, 6, 12, 24, 48, 96, 192 & 384, four is added, giving 4, 7, 10, 16, 28, 52, 100, 196 & 388. These numbers fit the planetary orbital radii really quite well (except for Neptune). The formula predicted a missing planet at 28 units between Mars and Jupiter and on 1st January 1801 Piazzi discovered Ceres, the largest of the asteroids in the asteroid belt, in the correct orbit.

The length of time it takes a planet to go once round the Sun is known as its *period*. Sometimes periods occur as simple ratios of each other, a famous example being the 2:5 ratio of Jupiter and Saturn (99.3%). Uranus, Neptune and tiny Pluto are especially rhythmic and harmonic, displaying a 1:2:3 ratio of periods, Uranus' and Neptune's adding to produce Pluto's (99.8%).

Like a whirlpool, inner planets orbit the Sun much faster than outer planets and the table (*opposite below*) shows the number of days between two planets' kisses, passes or near approaches, properly called *synods*. Does Earth experience any harmonics? Well, we have two planetary neighbours, Venus sunside and Mars spaceside and the figures reveal that we kiss Mars *three* times for every *four* Venus kisses (99.8%). So an ultraslow 3:4 rhythm or a deep musical fourth is being played around us *all* the time!

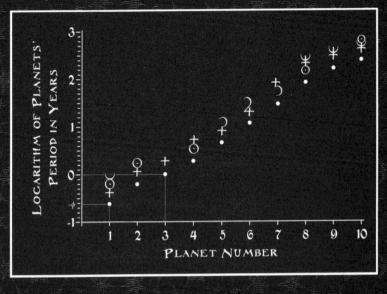

	☿	♀	+	♁	☽	♃	♄	♅	♆	♀
☿	∞	144.6	115.9	100.9	92.83	89.79	88.70	88.22	88.10	88.05
♀	144.6	∞	583.9	333.9	259.4	237.0	229.5	226.4	225.5	225.3
+	115.9	583.9	∞	779.9	466.7	398.9	378.1	369.7	367.5	366.7
♁	100.9	333.9	779.9	∞	1,162	816.5	733.9	702.7	694.9	692.2
☽	92.83	259.4	466.7	1,162	∞	2,744	1,991	1,777	1,728	1,712
♃	89.79	237.0	398.9	816.5	2,744	∞	7,252	5,045	4,669	4,551
♄	88.70	229.5	378.1	733.9	1,991	7,252	∞	16,570	13,100	12,210
♅	88.22	226.4	369.7	702.7	1,777	5,045	16,569	∞	62,890	46,440
♆	88.10	225.5	367.5	694.9	1,728	4,669	13,100	62,890	∞	179,800
♀	88.05	225.3	366.7	692.2	1,712	4,551	12,210	46,440	179,800	∞

THE INNER PLANETS
Mercury, Venus, Earth and Mars

Our solar system can be thought of a series of thin rotating rings, each slowly settling down. Divided by an asteroid belt into two halves, the inner region sports four small rocky planets quickly orbiting the Sun, while the outer half has four slow huge gas and ice planets.

The Sun has still not given up its secrets. Mostly Hydrogen and Helium, and an element factory, it is also a giant fluid geometric magnet, 15 million°C at its core, 6,000°C at the surface. It blows a particle wind through the entire solar system and its sunspots and huge solar flares affect electronics on Earth.

Mercury is the first planet. Mostly solid iron, it is a cratered, atmosphereless world, 400°C in the sunshine, -170°C in the shade.

Venus is second, a cloud-shrouded greenhouse world. On the surface the temperature is a staggering 480°C and the carbon-dioxide rich atmosphere is *ninety* times denser than Earth's. An apple here would be instantly incinerated by the heat, crushed by the atmosphere and finally dissolved in sulphuric acid rain.

Earth is the third planet, the one with life and one Moon.

Mars is fourth, a rocky red world, just above freezing. Ice caps cover the poles under a thin atmosphere. River beds suggest that Mars may once have had oceans but they are long gone now and today dust storms regularly envelop the planet for days. Huge dead volcanoes, one three times larger than Mount Everest, stand witness to a bygone age. Mars has two tiny moons.

Beyond Mars is the Asteroid Belt, and, beyond that, the giants.

SIZES OF THE INNER PLANETS

TILTS AND ECCENTRICITIES OF THE ORBITS OF THE INNER PLANETS

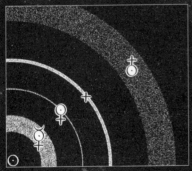

SUN-CENTRED

VIEW FROM EARTH

UNDERSTANDING THE PICTURES
a few tips on things which appear in this book

Seen from Earth, day or night, the Sun moves slowly to the left against the stars (right in the southern hemisphere), taking a year to return to the same star. The Moon swiftly circles around in the same direction every month, taking 27.3 days to return to a star. Venus and Mercury oscillate around the Sun, coming and going, as the Sun itself slowly trundles around its yearly circle. Imagine standing on Venus – the Sun moves faster against the stars and Mercury is closer, whirling round the Sun like a fairground waltzer.

Every *pair* of planets creates a *single* dance. It doesn't matter which of the two you stand on, your partner's dance around you will be the same. It is a shared experience. Mercury's evolving waltzes with Earth and Venus are shown (*opposite top*). Earth and Mercury roughly kiss 22 times in 7 years, though the ancient Greeks also knew of a more accurate 46 year, 145 synod cycle. Mercury and Venus are beautifully in tune after just 14 kisses.

Shown opposite below is the Golden Ratio, ϕ or *phi*. It is found throughout every pentagram and in the Fibonacci series of numbers (*shown opposite*), which starts with 1, 2, 3, 5, 8 and 13, all numbers we will see in the inner planets. The Golden Ratio is essentially 0.618, but since one *divided* by it is 1.618 (which is the same as *adding* one to it), and 1.618 *times* 1.618 equals 2.618 (the same as adding one more), it often takes all these values. The Golden Ratio is found throughout organic life forms; it is the signature of life, and, as we shall see, highly accurately present in the inner solar system too.

240 DAYS   770 DAYS   2030 DAYS

MERCURY'S AND VENUS' DANCE

470 DAYS   1390 DAYS   2510 DAYS

MERCURY'S AND EARTH'S DANCE

1	
0 + 1 = 1	1 ÷ 1 = 1
1 + 1 = 2	1 ÷ 2 = 0.5
1 + 2 = 3	2 ÷ 3 = 0.6667
2 + 3 = 5	3 ÷ 5 = 0.6
3 + 5 = 8	5 ÷ 8 = 0.625
5 + 8 = 13	8 ÷ 13 = 0.6154
8 + 13 = 21	13 ÷ 21 = 0.6190
13 + 21 = 34	21 ÷ 34 = 0.6176
21 + 34 = 55	34 ÷ 55 = 0.6182
34 + 55 = 89	55 ÷ 89 = 0.6180

1 2 3 5 8 13 ... ϕ = 0.61803399...

1.618   1.618

1

2.618

THE GOLDEN RATIO   FIBONACCI NUMBERS

MERCURY & VENUS' ORBITS
a very simple aide-memoire

There are few things more simple than a circle. With Kepler's discovery of the ellipses, and Newton and Einstein setting them spinning, the planetary orbits can be thought of as orbital 'circles', centred on the Sun, with the eccentricity thickening the circle slightly, or giving the spheres a shell (*see Kepler's diagram, page 13*).

One of the very first things you can do with circles is to put three of them together so that they all touch. Amazingly, the orbits of the first two planets of the solar system are hiding in this simple design. If Mercury's mean orbit passes through the centres of the three circles then Venus' encloses the figure (99.9%).

This is a simple trick to remember – you see it all around you all the time, in the home, in design, art, architecture and nature. Every time you pick up three glasses or push three balls together you create the first two planets' circular orbits, to an extraordinary degree of accuracy. There must be a reason for this beautiful fit between the ideal and the manifest, but none is yet known and these kinds of problems are currently out of fashion; perhaps a bright 21st century scientist will find an answer – until then it remains a 'coincidence'.

The triangle is one emblem of the musical octave 2:1, and Mercury performs a delightful solo on this theme, as one Mercury day is exactly two Mercury years, during which time the planet has spun on its own axis exactly three times. Thus the very first planet plays the very first harmonies and draws one of the very first geometrical shapes. We have started with one, heard a two, and seen a three.

THE KISS OF VENUS
our most beautiful relationship

Other than the Sun and Moon, the brightest point in the sky is Venus, morning and evening star. She is our closest neighbour, kissing us every 584 days as she passes between us and the Sun. Each time one of these kisses occurs the Sun, Venus and the Earth line up two-fifths of a circle further around the starry zodiacal circle - so a pentagram of conjunctions is drawn. Seen from Earth the Sun moves round the zodiac while Venus whirls around the Sun drawing an astonishing pattern over exactly eight years (99.9%), which is thirteen Venusian-years (99.9%). Small loops are made when Venus in her dazzling kiss seems briefly to reverse direction against the background Stars (*shown below as seen from Earth*). Notice the Fibonnacci numbers we have just met, 5, 8 and 13. The periods of Earth and Venus are also loosely related as 1.618:1 (99.6%).

This 'phi'-fold nature of Venus and Earth's dance extends to their closest and furthest distances from each other. Opposite we see Venus' *perigee* and *apogee* precisely defined by two pentagrams, 2.618:1 (99.99%). All these diagrams *also* apply to Venus' experience of Earth.

THE PERFECT BEAUTY OF VENUS
the things they don't teach you at school

With the Sun in the centre, let us look at the orbits of Venus and the Earth. Every couple of days a line is drawn between the two planets (*below left*). Because Venus orbits faster she completes a whole circuit in the same time that the Earth completes just over a half-circuit (*below centre*). If we keep watching for exactly eight years the pattern opposite emerges, the sun-centred version of the five-petalled flower on the previous page.

The ratio between Earth's outer orbit and Venus' inner orbit, i.e their *home*, is intriguingly given by a square (*below right*) (99.9%).

Venus rotates extremely slowly on her own axis in the opposite direction to most rotations in the solar system. Her rotation period is precisely two-thirds of an Earth year, a musical fifth. This closely harmonises with the dance opposite so that every time Venus and Earth kiss, Venus does so with her *same face*. So paint a spot on Venus' surface and every time she lines up with the as seen from Earth the spot will point back at Earth. Over the eight Earth years of the five kisses, Venus spins on her own axis twelve times in thirteen of her years (*from Kollerstrom*). All beautiful musical numbers.

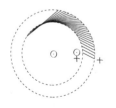  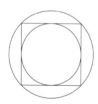

MERCURY & EARTH
yet more phives and eights

Mercury and Earth's physical sizes are in the same relation as their mean orbits! Various five and eight-fold overlays are shown opposite which proportion the orbits *and* sizes of the two planets.

The diameter of Mercury's *innermost* orbit is suggested by the pentagram incircle (*top left*) (99.5%) and also happens to be the distance between the mean orbits of the two planets (99.7%).

The diagram bottom right expands on the three touching circles of page 21. Eight circles centred on Venus' orbit produce Earth's mean orbit (99.99%) – the eight years of the five kisses perhaps?

Mercury, Venus and Earth display a peculiar coincidence: If we work in units of Mercury's orbital radius and period, then Venus' period times 2.618 is Earth's orbital radius squared (99.8%). Mercury's dance around Earth also produces its synodic year of 115.9 days. Richard Heath recently discovered that this is 2.618 times a musical fifth times a full moon (99.9%) (a musical fifth is 3:2; 2.618 is ϕ^2, or 1.618 x 1.618 and there is a full moon every 29.53 days).

Earth's and Saturn's relative orbits *and* sizes are given by a fifteen-pointed star (*below*), which also produces the tilt of the Earth.

THE RELATIVE PHYSICAL SIZES OF MERCURY AND EARTH CAN
BE DEFINED BY EITHER A PENTAGRAM OR AN OCTAGRAM (99%)

THE RELATIVE ORBITS OF MERCURY AND EARTH CAN ALSO BE
DEFINED BY EITHER A PENTAGRAM OR OCTAGRAM (99%)

OTHER MORE ACCURATE WAYS OF DRAWING THE INNER ORBITS (99.9%)

THE ALCHEMICAL WEDDING
three to eleven all round

From the surface of the Earth, the Sun and the Moon appear the same size. According to modern muggle cosmology this is 'just' a coincidence, but any good wizard will tell you the balance between these two primary bodies is clear proof of very ancient magic.

The size of the Moon compared to the Earth is 3 to 11 (99.9%). What this means is that if you draw down the Moon to the Earth, then the circle through the centre of the heavenly Moon will have a circumference equal to the perimeter of a square enclosing the Earth. The ancients seem to have known about this, and hidden it in the definition of the mile (*opposite, after Michell & Ward*).

The Earth-Moon proportion is also precisely invoked by our two neighbours, Venus and Mars (*Venus shown dancing round Mars below*). The closest:farthest distance ratio that each experiences of the other is, incredibly, 3:11 (99.9%). The Earth and the Moon sit in between them, perfectly echoing this beautiful local spacial ratio.

3:11 happens to be 27.3% and the Moon orbits the Earth every 27.3 days, the same period as the average rotation period of a sunspot. The Sun and Moon do seem very much the unified couple.

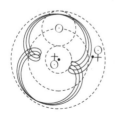 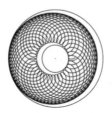 

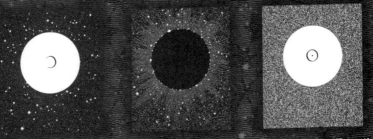

THE MOON, A TOTAL SOLAR ECLIPSE AND THE SUN, AS SEEN FROM EARTH

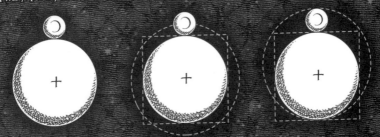

THE SIZES OF THE MOON AND THE EARTH 'SQUARE THE CIRCLE'
THE DASHED SQUARE AND CIRCLE ARE THE SAME LENGTH OF STRING

MILES OF MOON AND EARTH

RADIUS OF MOON = 1080 MILES = 3 × 360 MILES

RADIUS OF EARTH = 3960 MILES = 11 × 360 MILES

DIAMETER OF MOON = 2160 MILES = 3 × 1 × 2 × 3 × 4 × 5 × 6 MILES

RADIUS OF EARTH + RADIUS OF MOON = 5040 MILES

= 1 × 2 × 3 × 4 × 5 × 6 × 7 = 7 × 8 × 9 × 10 MILES

DIAMETER OF EARTH = 7920 MILES = 8 × 9 × 10 × 11 MILES

THERE ARE 5280 FEET IN A MILE

= (10 × 11 × 12 × 13) - (9 × 10 × 11 × 12)

CALENDAR MAGIC
just three numbers do the trick

Recent work by Robin Heath has revealed simple geometrical and mathematical tools which suggest order and form within the Sun-Moon-Earth system. Imagine we want to discover the number of full moons in a year (somewhere between 12 and 13). Draw a circle, diameter thirteen with a pentagram inside. Its arms will then measure 12.364, the number of full moons in a year (99.95%).

An even more accurate way of doing it is to draw the second Pythagorean triangle, which just happens to be made of 5, 12 and 13 again, the numbers of the keyboard, and of Venus (*page 26*). Dividing the 5 side into its harmonic 2:3 gives a new length, the square root of 153, 12.369, the number of full moons in a year (99.999%).

The Moon seems to beckon us to look further. We all know that six circles fit around one on a flat surface (6 and 7). Twelve spheres pack perfectly around one in our familiar three-dimensional space (12 and 13 again). We seem to be moving up in sixes. Could *eighteen* time-spheres fit around one in a fourth dimension of time? Incredibly, all of the current major time cycles of the Sun-Moon-Earth system turn out to be accurately defined as simple combinations of the numbers 18, 19 and the Golden Section (*see page 22*).

The Golden Section is evident in the pentagram, the icosahedron, the dodecahedron and all living things. The orbits of the four inner planets all display its presence. Its values 0.618, 1, 1.618 and 2.618 added to the magic number 18 produce 18, 18.618, 19, 19.618 and 20.618, which then multiply together as shown opposite.

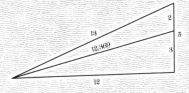

TWO ANCIENT TECHNIQUES
FOR FINDING THE NUMBER
OF FULL MOONS IN A YEAR

18 YEARS = THE SAROS ECLIPSE CYCLE (99.83%)
(SIMILAR ECLIPSES WILL OCCUR AFTER 18 YEARS)

18.618 YEARS = REVOLUTION OF THE MOON'S NODES (99.99%)
(THE MOON'S NODES ARE THE TWO PLACES WHERE THE SLIGHTLY
OFFSET CIRCLES OF THE SUN AND MOON'S ORBITS CROSS)

19 YEARS = THE METONIC CYCLE (99.99%)
(IF THERE IS A FULL MOON ON YOUR BIRTHDAY THIS YEAR - THERE
WILL BE ANOTHER ONE ON YOUR BIRTHDAY IN 19 YEARS TIME)

THE ECLIPSE YEAR = 18.618 X 18.618 DAYS (99.99%)
(THE ECLIPSE YEAR IS THE TIME IT TAKES FOR THE SUN TO RETURN
TO THE SAME ONE OF THE MOON'S NODES IT IS 18.618 DAYS SHORT
OF A SOLAR YEAR (99.99%). THERE ARE 19 ECLIPSE YEARS IN A SAROS)

12 FULL MOONS = 18.618 X 19 DAYS (99.82%)
(12 FULL MOONS IS THE LUNAR OR ISLAMIC YEAR)

THE SOLAR YEAR = 18.618 X 19.618 DAYS (99.99%)
(THE SOLAR YEAR IS THE 365.242 DAY YEAR WE ARE USED TO)

13 FULL MOONS = 18.618 X 20.618 DAYS (99.99%)
(13 FULL MOONS IS ANOTHER 18.618 DAYS AFTER THE SOLAR YEAR)

COSMIC FOOTBALL
Mars, Earth and Venus spaced

The next planet out from Earth is the fourth planet, Mars. Kepler had tried a *dodecahedron* spacing the orbits of Mars and Earth and an *icosahedron* spacing Earth from Venus (*see page 12*), and, coincidentally, it turns out he was very close to the mark.

The dodecahedron (made of twelve pentagons) and the icosahedron (made of of twenty equilateral triangles) are the last two of the five perfect polyhedra (the *Platonic Solids*). They form a pair, as each creates the other from the centres of its faces (*below*). Opposite, they appear in bubble form inside Mars' spherical orbit. The dodecahedron magically produces Venus' orbit as the bubble within (*opposite top*) (99.98%), while the icosahedron defines Earth's orbit through its bubble centres (*lower opposite*) (99.9%).

In the ancient sciences the icosahedron was associated with the element of water, so it is appropriate to see it emanating from our watery planet. The dodecahedron represented *aether*, the life force, here enveloping lively Earth, and defined by her two neighbours.

VENUS AND
MARS

EARTH
AND MARS

THE ASTEROID BELT
through the looking glass

We have reached the end of the inner solar system. Beyond Mars lies a particularly huge space, the other side of which is the enormous planet Jupiter. It is in this space that the Asteroid Belt is found, thousands of large and small tumbling rocks, silicaceous, metallic, carbonaceous and others. There are spaces, *Kirkwood Gaps*, in the asteroid belt, cleared where orbital resonances with Jupiter occur. The largest gap is at the orbital distance which would correspond to a period of one third that of Jupiter.

The largest of the asteroids by a very long way is Ceres, comprising over one third of the total mass of all of them. She is about the size of the British Isles and produces a *perfect* eighteen-fold pattern with Earth (*see page 56*).

Bode's Law predicted something at the distance of the asteroid belt (*see page 16*), but it was Alex Geddes who recently discovered the weird mathematical relationship between the four small inner planets and the four outer gas giants. Their orbital radii magically 'reflect' about the asteroid belt and multiply as shown below and opposite to produce two enigmatic constants.

$$Ve \times Ur = 1.204 \; Me \times Ne \qquad Ve \times Ma = 2.872 \; Me \times Ea$$
$$Me \times Ne = 1.208 \; Ea \times Sa \qquad Sa \times Ne = 2.876 \; Ju \times Ur$$
$$Ea \times Sa = 1.206 \; Ma \times Ju \qquad (Ve \times Ma \times Ju \times Ur = Me \times Ea \times Sa \times Ne)$$

The asteroid belt is unlikely to be the remains of a small planet as no sizeable body could ever have formed so close to Jupiter.

THE ASTEROID BELT SEPARATES THE FOUR SMALL ROCKY
INNER PLANETS FROM THE FOUR HUGE OUTER ONES

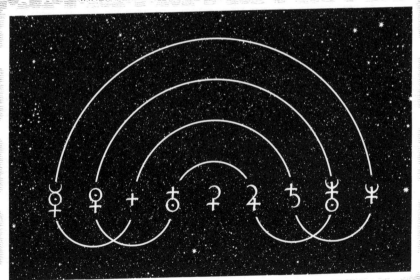

THE PATTERN OF GEDDES' MAGICAL MULTIPLICATIONS

THE OUTER PLANETS
Jupiter, Saturn, Uranus, Neptune & beyond

Beyond the Asteroid Belt we come to the realm of the gas and ice giants, Jupiter, Saturn, Uranus and Neptune.

Jupiter is the largest planet, and its magnetic field is the largest object in the solar system. 90% hydrogen, it is nevertheless built around a rocky core like all the giant planets. Metallic hydrogen and then liquid hydrogen surrounds this core. The famous Red Spot is a storm, larger than Earth, which has raged now for hundreds of years. Jupiter's moons are numerous and fascinating: One, Io, is the most volcanic body in the solar system; another, Europa, may have warm oceans of water beneath its icy surface.

The next planet is Saturn, with its beautiful system of rings. Saturn's structure beneath its clouds is much the same hydrogen and helium mix as Jupiter. A large number of moons have been discovered, the largest of which is Titan, a world the size of Mercury with all the building blocks for life.

Beyond Saturn is Uranus, which orbits on its side. Winds gust on the equator at six thousand times the speed of sound.

Next is Neptune, like Uranus an ice world of water, ammonia and methane. The largest moon, Triton, has nitrogen ice caps and geysers which spew liquid nitrogen high into the atmosphere.

Finally, tiny Pluto, and, beyond that, the primordial swarm of the Kuiper Belt. Then, stretching a third of the way to the nearest star, the sphere of icy debris and comets of the Oort Cloud.

SIZES OF THE OUTER PLANETS

TILTS AND ECCENTRICITIES OF THE ORBITS OF THE OUTER PLANETS

SUN-CENTRED

VIEW FROM THE EARTH

FOURS

Mars, Jupiter and massive moons

An asteroid belt and 550 million km separate Mars' and Jupiter's orbits, further than Earth is from the Sun. Jupiter is the first and largest of the gas giants, the vacuum cleaner of the solar system. If Jupiter had gathered only slightly more material during its long and ongoing formation its internal pressures would have turned it into a star and we would have had a second Sun.

The top diagram opposite shows a really simple way to draw the orbits of Mars and Jupiter from four touching circles or a square (99.98%). It is a proportion commonly seen in church windows and railway stations. Shown below, on this page, is a pattern from the same family, which spaces Earth's and Mars' orbits (99.9%).

Jupiter has four particularly large moons. The two largest, Ganymede and Callisto, are the size of the planet Mercury and produce one of the most perfect space-time patterns in the solar system. An observer living on either moon would experience the motions of the other in space and time as the beautiful fourfold diagram shown opposite.

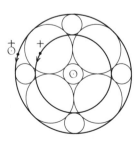

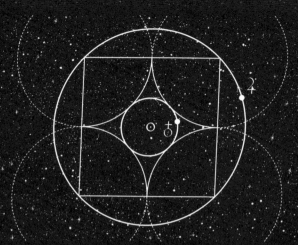

HOW TO ACCURATELY DRAW MARS & JUPITER'S MEAN ORBITS

THE BEAUTIFUL DANCE OF GANYMEDE & CALLISTO

OUTER MOONS
harmonic patterns

Four groups of moons orbit Jupiter. The first two groups each have four moons so look very like a model of the whole solar system – four small inner bodies followed by four giants. The second group, of four particularly large moons, the *Galileans*, is further divided into two small rocky worlds, Io and Europa, then two gas and ice moons the size of planets, Ganymede and Callisto.

The grouping into fours is very striking indeed. Each of the four groups has its own general moonsize, orbital plane, period and distance from Jupiter (the inclinations of the four orbital planes of the four groups even add up to a quarter of a circle (99.9%)).

Saturn has over thirty moons, most shepherding and tuning the amazing rings with the larger bodies tending to be further out. Far beyond Saturn's rings, however, are three moons – the gigantic Titan, tiny Hyperion and, further out still, Iapetus.

Opposite we see fascinating harmonic patterns: two from Jupiter's largest moons, two experienced by Saturn's giant moon Titan and two experienced by Neptune, the outer sizeable planet of the solar system. Disharmony would not make pretty pictures.

MERCURY · THE MOON · IO · EUROPA · GANYMEDE · CALLISTO · TITAN · PLUTO

EUROPA & IO

EUROPA & GANYMEDE

TITAN & HYPERION

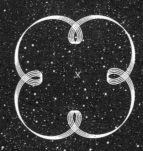

TITAN & IAPETUS

URANUS & NEPTUNE

NEPTUNE & PLUTO

JUPITER'S GIANT SEAL
huge hexagrams and affirmatory asteroids

Jupiter, the largest planet, was king of the ancient gods, Zeus to the Greeks. A delightful feature of its orbit is its associated pair of asteroid clusters. *The Trojans* are two groups of asteroids which move round Jupiter's orbit, 60° ahead of it and 60° behind (*opposite*). This partnership perpetually moves round the Sun as though held in place by the spokes of a wheel. The positions of the Trojan clusters are known as the *Laplace Points*, with Sun, Jupiter and Trojans forming gravitationally balanced equilateral triangles.

Just for the fun of it, if we now join the spokes as shown opposite then three hexagrams can be seen to produce Earth's mean orbit from Jupiter's (99.8%), – a very easy trick to remember. Earth and Jupiter's orbits are thus lurking in every crystal. Another name for a six-pointed star made of two triangles is a *Star of David* or *Seal of Solomon*, a kingly design indeed, by Jove!

Exactly the same Earth-Jupiter proportion may be created by spherically nesting three cubes, or three octahedra, or any threefold combination of them (*two are shown alternating below*).

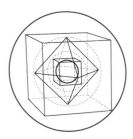
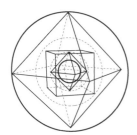

THE GOLDEN CLOCK
Jupiter and Saturn seen from Earth

Jupiter and Saturn are the two largest planets of the solar system and ruled the outer two spheres of the ancient system. In ancient mythology, Saturn was Chronos, the Lord of Time.

The top two diagrams opposite show the close 5:2 ratio of their periods. Top left we see their dance; the beautiful three-fold harmonic is immediately apparent, spinning slowly because of the slight miss in the harmony. From Earth, this pattern is seen as an important sequence of conjunctions and oppositions of Jupiter and Saturn, who kiss every 20 years. Top right we see the hexagram created by these positions – with conjunctions marked on the outside of the zodiac and oppositions marked inside. The planets move anticlockwise around the dashed circle of the ecliptic, starting at twelve o'clock, Jupiter moving faster than Saturn.

The lower diagram shows the relative speeds of orbit of Earth, Jupiter and Saturn. We start with the three planets in a synodic line at twelve o'clock. Earth orbits much faster than the outer planets and makes a complete circuit of the Sun (365.242 days) and then a bit more before lining up with slowcoach Saturn again for a synod after 378.1 days. Three weeks later it lines up again with Jupiter (after 398.9 days). Richard Heath recently discovered that the Golden Ratio is defined here in time and space to a *stunning* 99.99% accuracy! The two giants of our solar system thus reinforce life on Earth.

Amusingly, Saturn takes the same number of years to go round the Sun as there are days between full Moons (99.8%)

JUPITER & SATURN'S DANCE

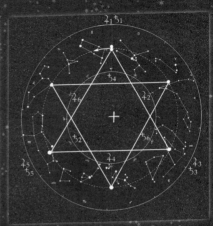

CONJUNCTIONS & OPPOSITIONS

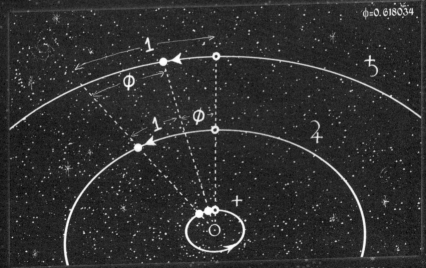

φ=0.618034

JUPITER & SATURN'S SYNODS DEFINE THE GOLDEN SECTION

OCTAVES OUT THERE
threes and eights again

If you ever want to incorporate Jupiter, Saturn and Uranus' orbits into a window or floor design the diagram opposite might help. An equilateral triangle and an octagram proportion the outer, mean and inner orbits of the three largest planets, like a spikey inversion of the touching circles solutions for the first three planets (*page 29*). Tiny inaccuracies can be seen, but the scheme is most memorable. An equilateral triangle defines two circles in an octave or 2:1 ratio.

Another rule of thumb is to remember that if Jupiter's orbit is 6, then Saturn's is 11 (99.9%), twice the Moon:Earth size ratio (*page 30*). Saturn's mean orbit also happens to invoke π or '*pi*' – twice: Its radius is the circumference of Mars' mean orbit (99.9%) and its circumference is the diameter of Neptune's mean orbit (99.9%).

Uranus and Neptune, like Saturn, both have ring systems with clear spaces at Kirkwood distances where particles orbit at periods harmonic with one or more moons. Uranus' bright outer ring has a diameter twice that of Uranus itself (99.9%), echoing the orbits of Uranus and Saturn, and Neptune's innermost ring is two-thirds the size of its outermost (99.9%). These proportions beautifully invoke the local timing, as Neptune's orbital period is twice that of Uranus, and Uranus' is two-thirds that of Pluto, an outer reflection of the inner harmonic 1:2:3 we saw with Mercury.

The Milky Way, the plane of our own galaxy, is tilted at almost exactly 60° to the ecliptic or plane of our solar system (99.7%). Every year the Sun crosses the galaxy through the galactic centre, and being alive in these times means this happens every midwinter's day.

ICE HALOS
rainbows where planets lie

Occasionally, if you look, you will see a pair of rainbow spots left and right of the Sun. Known as 'Sun dogs', these are the first elements to appear of an ice halo—a thin rainbow circle around the Sun. Caused by light passing through ice crystals high in the atmosphere, Sun dogs appear 22.5° left and right of the Sun, just outside the bright 22° halo. Sometimes, a second larger halo appears 46° from the Sun, with a distinctive arc on top, the whole arrangement looking strangely similar to the ancient glyph for Mercury.

Amazingly, these two ice halos match the mean orbits of the inner two planets Mercury and Venus seen from the surface of the Earth. So, when you look at a double ice halo, you really are seeing the spheres of the mean orbits of Mercury and Venus, hanging in the sky. And what is more, the same two ice halos also function as a diagram of the relative orbits of Venus and Mars.

This is extraordinary. Every circle fits. Sunlight and ice dust paint orbits as rainbows while the Sun and Moon appear the same size in the sky. Our closest neighbour dances 5-fold around us in 8 years or 13 of her years, while below on Earth plants also dance 5, 8 and 13. These coincidences are focused on us, here, and now, a planet of conscious observers. Have we played some part in this. Could the fact of observation have lensed reality in some way?

Do beautiful and improbable coincidences surround other planets of observers? The suggestion is that they do - we may in fact be living in a conscious quantum holographic universe.

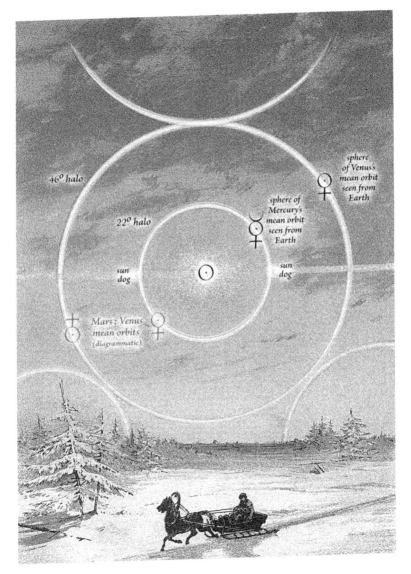

46° halo

22° halo

sun dog

sun dog

sphere of Mercury's mean orbit seen from Earth

sphere of Venus's mean orbit seen from Earth

Mars : Venus mean orbits (diagrammatic)

THE STARRY SIGNATURE
circumstantial evidence for life on earth

Despite all the scientific discoveries over recent centuries we are possibly today as far from understanding what we are doing here as the ancients were from being able to build a pocket calculator. The ancients, however, pondered consciousness deeply, and held that the soul was particularly akin to the applied arts of geometry and music. Through these arts they carefully investigated the relationship between 'the One' and 'the Few', for in music there are only so many notes in tune, and in geometry only so many shapes that fit. In recent centuries, the equations of Newton, Einstein and others have further refined our understanding towards simplicity and beauty.

This book has shown simple and beautiful examples of harmony and geometry in the solar system. The Golden Ratio, long associated with life, and conspicuously absent from modern equations, plays especially lovingly around Earth. Does this in some way have something to do with 'why we are here', and if so could these techniques be used to locate intelligent life in other solar systems?

If you ever need reminding that there may be a little more magic to our origins than modern cosmology can yet offer, then just remember the kiss of Venus and the words of John Donne:

> *"Man hath weav'd out a net, and this net throwne*
> *upon the Heavens, and now they are his owne.*
> *Loth to goe up the Hill, or labour thus*
> *to goe to Heaven, we make Heaven come to us."*

SUN & PLANETS	Perihelion $(10^6$ km$)$	Mean Orbit $(10^6$ km$)$	Aphelion $(10^6$ km$)$	Eccentricity	Inclination of Orbit (degrees)	Perihelion Longitude (degrees)	Orbital Period (days)	Tropical Year (days)
The Sun ☉	-	-	-	-	-	-	-	-
Mercury ☿	46.00	57.91	69.82	0.205631	7.0049	77.456	87.969	87.968
Venus ♀	107.48	108.21	108.94	0.006773	3.3947	131.53	224.701	224.695
The Earth ⊕	147.09	149.60	152.10	0.016710	0	102.95	365.256	365.242
Mars ♂	206.62	227.92	249.23	0.093412	1.8506	336.04	686.980	686.973
Ceres ⚳	446.60	413.94	381.28	0.0789	10.58	???	1680.1	1679.5
Jupiter ♃	740.52	778.57	816.62	0.048393	1.3053	14.753	4,332.6	4,330.6
Saturn ♄	1,352.2	1,433.5	1,514.5	0.054151	2.4845	92.432	10,759.2	10,746.9
Chiron ⚷	1,266.2	2,050.1	2,833.9	0.38316	6.9352	339.58	18,518	18,512
Uranus ♅	2,741.3	2,872.46	3,003.6	0.047168	0.76986	170.96	30,685	30,589
Neptune ♆	4,444.4	4,495.1	4,545.7	0.0085859	1.7692	44.971	60,190	59,800
Pluto ♇	4,435.0	5,869.7	7,304.3	0.24881	17.142	224.07	90,465	90,588

MOONS (a selection)	Name of Satellite	Mean Orbital Radius $(10^3$ km$)$	Orbital Period (days)	Eccentricity of Orbit	Inclination of Orbit (°)	Diameter (mean) (km)	Mass $(10^{18}$ kg$)$
Earth's ⊕	The Moon	384.8	27.3217	0.0549	5.145	3,475	73,490
Mars' ♂	Phobos	9,378	0.31891	0.0151	1.08	22.4	0.0106
	Deimos	23,459	1.26244	0.0005	1.79	12.2	0.0024
Jupiter's ♃	Io	421.6	1.7691	0.004	0.04	3,643	89,330
	Europa	670.9	3.5512	0.009	0.47	3,130	47,970
	Ganymede	1,070	7.1546	0.002	0.21	5,268	148,200
	Callisto	1,883	16.689	0.007	0.51	4,806	107,600
Saturn's ♄	Tethys	294.66	1.8878	<0.001	1.86	1,060	622
	Dione	377.40	2.7369	0.0022	0.02	1,120	1,100
	Rhea	527.04	4.5175	0.0010	0.35	1,528	2,310
	Titan	1,221.8	15.945	0.33	0.33	5,150	134,550
	Iapetus	3,561.3	79.330	0.0283	14.7	1,436	1,590

Rotation Period (hours)	Average Day Length (hours)	Equatorial Diameter (km)	Polar Diameter (km)	Axial Tilt (degrees)	Mass (10^{24} kg)	Volume (10^{12} km³)	Surface Gravity (m/s²)	Surface Pressure (bars)	Temp. (mean) (°C)
600 – 816	–	1,392,000	1,392,000	7.25	1,989,100	1,412,000	274.0	0.000868	5505
1407.6	4222.6	4,879.4	4,879.4	0.01	0.3302	0.06083	3.70	negl.	167
-5832.5	2802.0	12,103.6	12,103.6	177.36	4.8685	0.92843	8.87	92	464
23.934	24.000	12,756.2	12,713.6	23.45	5.9736	1.08321	9.78	1.014	15
24.623	24.660	6794	6750	25.19	0.64185	0.16318	3.69	0.007	-65
9.0744	9.0864	960	932	var.	0.00087	0.000443	negl.	negl.	-90
9.9250	9.9259	142,984	133,708	3.13	1,898.6	1,431.28	23.12	100+	-110
10.656	10.656	120,536	108,728	26.73	568.46	827.13	8.96	100+	-140
5.8992	5.8992	208	148	???	0.000006	0.000024	negl.	negl.	???
-17.239	17.239	51,118	49,946	97.77	86.832	68.33	8.69	100+	-195
16.11	16.11	49,528	48,682	28.32	102.43	62.54	11.00	100+	-215
-153.29	153.28	2390	2390	122.53	0.0125	0.00715	0.58	negl.	-223

MOONS (continued)		Name of Satellite	Mean Radius Orbit (10^3 km)	Orbital Period (days)	Eccentricity of Orbit	Inclination of Orbit (°)	Diameter (mean) (km)	Mass (10^{18} kg)
Uranus'	♅	Miranda	129.39	1.4135	0.0027	4.22	235.7	66
		Ariel	191.02	2.5204	0.0034	0.31	578.9	1,340
		Umbriel	266.30	4.1442	0.0050	0.36	584.7	1,170
		Titania	435.91	8.7059	0.0022	0.14	788.9	3,520
		Oberon	583.52	13.463	0.0008	0.10	761.4	3,010
Neptune's	♆	Proteus	117.65	1.1223	0.0004	0.55	193	3
		Triton	354.76	-5.8769	0.000016	157.35	2,705	21,470
		Nereid	5,5413	360.14	0.7512	7.23	340	20
Pluto's	♀	Charon	19.6	6.3873	<0.001	<0.01	1,186	1,900

Only the major moons of the gas giants are given. In 2001 there were 28 known moons around Jupiter, 30 around Saturn, 21 around Uranus and 8 around Neptune. There are probably many more. There are 29.5306 days between full moons on Earth. Cosmology can seriously improve your health.

MERCURY - VENUS

MERCURY - EARTH

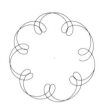

MERCURY - MARS

MERCURY - CERES

MERCURY - JUPITER

MERCURY - SATURN

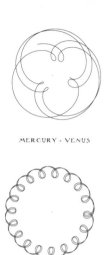

VENUS - EARTH

VENUS - MARS

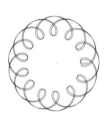

VENUS - CERES

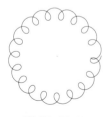

VENUS - JUPITER

VENUS - SATURN

EARTH - MARS